W0099651

Social Media Managers

Tamara Wilburn

Series Editor **Casey Malarcher**

Level 1 - 4

Social Media Managers

Tamara Wilburn

© 2018 Seed Learning, Inc.

All rights reserved. No part of this book may be reproduced, stored in a retrieval system, or transmitted in any form by any means, electronic, mechanical, photocopying, recording, or otherwise, without prior permission in writing from the publisher.

Series Editor: Casey Malarcher
Acquisitions Editor: Anne Taylor
Copy Editor: Liana Robinson
Cover/Interior Design: Highline Studio

ISBN: 978-1-943980-36-9

10 9 8 7 6 5 4 3 2 1
22 21 20 19 18

Photo Credits

All photos are © Shutterstock, Inc.

Contents

What Is a Social Media Manager?

Talking with people is important for a business. Today, an easy way to talk to people is on the internet. A business needs people to act as the business' voice over the internet.

These people are social media managers.

Talking with someone over the internet

Over half of the world has some way to go online.

They have computers.

They have smartphones.

And most of these people use social media.

◀ Many people have both a computer and a smartphone.

Social media apps

How often do you look at social media?
People who enjoy using social media are on the
websites day and night.
Someone you know might be using the internet to
check a social media website right now.

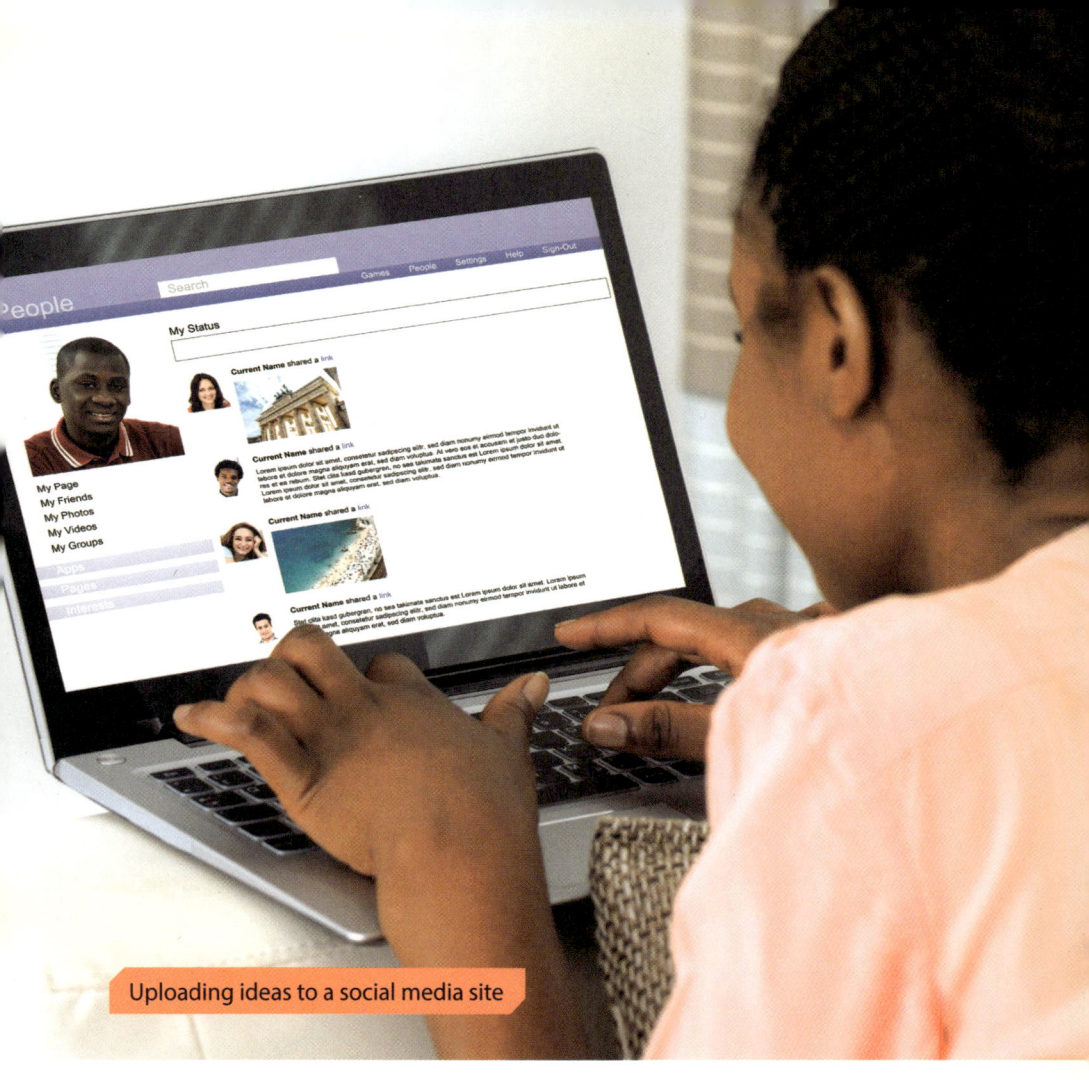

Uploading ideas to a social media site

Some social media websites are for connecting people.
Some websites are for passing along good ideas such as
how to cook or make things.
And many of these websites have lots of users.

Social media websites are important.

Users share their own thoughts, pictures, and ideas over the internet.

People can also interact with others.

These websites have changed how people interact in the 21st century.

Taking a picture to upload

Have you ever interacted with a business on social media?

Did you write to the business to ask a question?

If you did, a social media manager probably read your question.

Maybe he or she wrote back to you.

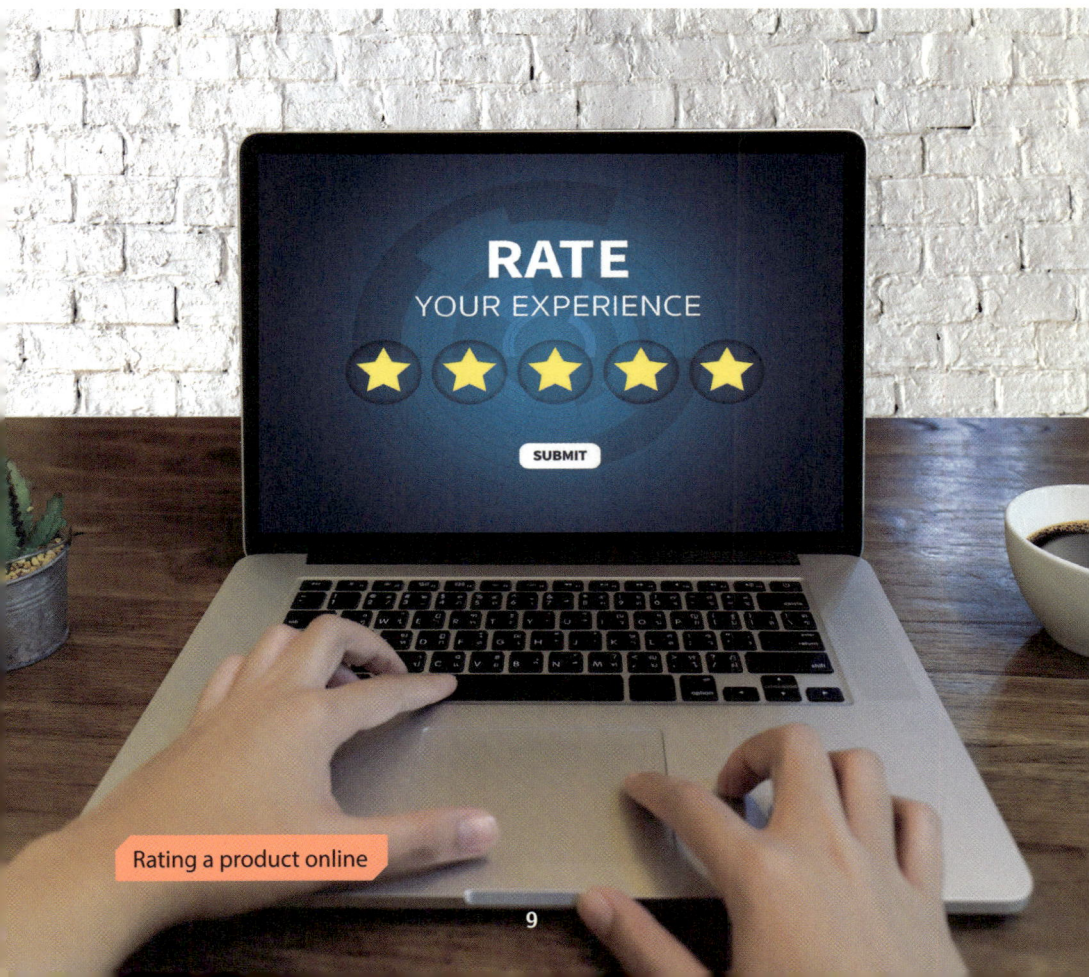

Rating a product online

Different kinds of people

A social media manager is the voice of a business on the internet.

A social media manager needs to know how to talk to the people attracted to the company he or she works for.

Businesses that attract younger people want to seem young.

The social media managers for these businesses need to know about things young people like.

They also need to use the kind of language that young people use.

This helps young people understand the business.

◀ Knowing how to respond to young people

Reading online reviews

Social media managers do more than talk with people. They also watch a business's reputation on the internet.

Social media managers know when people say good or bad things about the business they work for.
They can learn what people do or do not like.
This information is important to businesses.

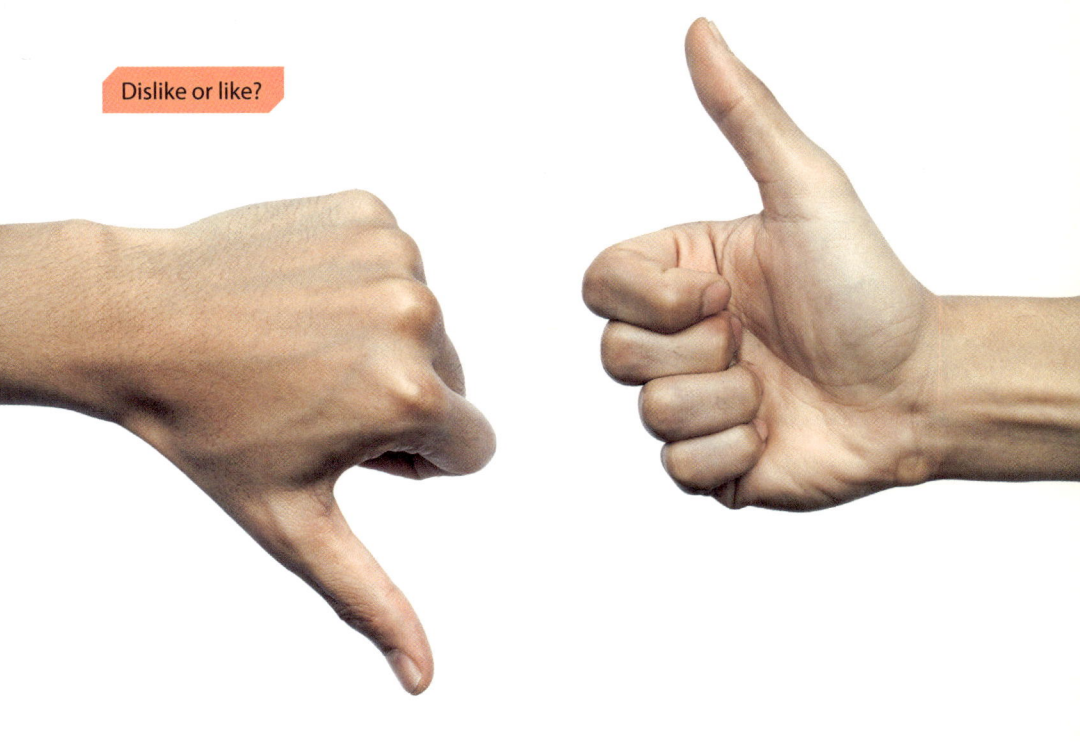

Dislike or like?

Social media managers work with marketing managers.
The marketing team creates internet advertising
campaigns.
Many of these campaigns change quickly.
Social media managers watch what people say about
these campaigns on the internet.

◄ Keeping an eye on
customer feedback

Sharing information with coworkers

Then social media managers give this information to the marketing managers.

The marketing managers use this information.

The information can tell them how to change an advertising campaign.

This can help make a campaign better.

Skills for Social Media Managers

Social media managers need to go to university. They should study computers, speaking, or marketing. A good social media manager will need all of these skills.

◀ A student with a university degree

Comfortable working on a computer

Social media managers use computers and the internet every day.

It's important for them to understand computers well.

Social media managers must communicate well. They will type most of their communication. They must have good writing skills and use correct grammar.

Writing is an important skill.

Knowing how businesses sell things is important, too.

Most businesses want to sell something to people.

A social media manager is like a person who sells something.

He or she is selling the ideas people should have about a business.

Making customers feel welcome ▶

19

Being the Voice of a Business

Social media managers must make quick decisions.
They must answer people in real time.
People like businesses that get back to them quickly.

◀ Getting a quick response online

◀ Upset by
a response

But social media managers also need to be careful.

It's easy to say something wrong.

This can cause trouble.

Talking together at the office

Social media managers must know what is best for a business.
They must talk with the people who work at the business with them all the time.

Seeing bad comments online

A social media manager needs to watch the company's website.

People may say bad things that cause trouble for the business.

But a good social media manager knows how to turn the bad to good.

The Future of Social Media

The internet has not been around for a long time.

It is new, and it changes every day.

This makes the internet a difficult place to do business.

Social media managers have to change with the internet.

Looking to the future

What works for a social media manager today might not work tomorrow.

In this job, people often must try new things no one has tried before when problems come up.

Finding new ways to interact with customers

CUSTOMER ENGAGEMENT

The internet will continue to change.

Social media managers' jobs will continue to change, too.

It's important to look for what is new and to keep learning.

This is an exciting time to work over the internet.

A new idea ▶

Do you want an interesting job that is always changing?

Are you interested in social media?

Then a job as a social media manager might be right for your future.

Comprehension Questions

1. Which of the following is a form of social media?
   (a) A computer
   (b) A smartphone
   (c) An app for cooking
   (d) All of the above

2. What can people do on social media?
   (a) Put their thoughts on the internet
   (b) Put pictures on the internet
   (c) Interact with others
   (d) All of the above

3. Social media managers know if people say _____ about the business they work for.
   (a) lies
   (b) good or bad things
   (c) jokes
   (d) secrets and riddles

4. What should social media managers study at university?
   (a) Computers
   (b) Speaking
   (c) Marketing
   (d) All of the above

5. Which of the following is NOT true about the internet?
   (a) It has been around for a long time.
   (b) It is changing.
   (c) It is a difficult place to do business.
   (d) It is used by both young and old people.

Key 1. (c) 2. (d) 3. (b) 4. (d) 5. (a)

Glossary

■ **advertising campaign** (n. phr.) an organized series of events to let people know about things to do or buy

■ **attract** (v.) to draw something toward another

■ **century** (n.) a period of 100 years

■ **customer** (n.) a person who is interested in buying something

■ **decision** (n.) the conclusion of a choice

- **interact** (v.) to talk or do things with other people

- **manager** (n.) a person whose job is to lead employees or maintain programs

- **marketing** (n.) the promoting and selling of products

- **reputation** (n.) the opinion that people have about what someone or something is like, based on what has happened in the past

- **website** (n.) a page or pages on the internet set up by a person or business to show information and pictures about the website's topic or purpose

Notes

Here are some companies known for using social media well. Readers may enjoy researching the social media programs of these companies to learn more about how businesses use social media.

Denny's Restaurant is a large chain restaurant. They put up strange, out of the ordinary things that people find entertaining on their social media page. Many people follow Denny's just to see what they'll put up next.

Dove has a beautiful goal when it comes to social media. The company wants to help women feel good about themselves. Dove is constantly creating content such as videos and campaigns aimed at showing that everyone is beautiful.

Pizza Hut likes to use humor on its social media website. The company retweets funny comments from customers, posts jokes, and tries to attract customers through humor.

The Walt Disney Company invites people to "share their ears" by posting a picture of themselves wearing the iconic Disney mouse ears. Smiling faces from all over have been posted on its social media website proving that people around the world love Disney.

List of Books

LEVEL 1

1. Robotics Engineers
2. Cyber Security Experts
3. Medical Scientists
4. Social Media Managers
5. Asset Managers

LEVEL 2

1. Drone Pilots
2. App Developers
3. Wearable Technology Creators
4. Computer Intelligence Engineers
5. Digital Modelers

LEVEL 3

1. IoT Marketing Specialists
2. Space Pilots
3. Water Harvesters
4. Genetic Counselors
5. Data Miners

LEVEL 4

1. Database Administrators
2. Nanotechnology Research Scientists
3. Quantum Computer Scientists
4. Agricultural Engineers
5. Intellectual Property Lawyers

"The future of the economy is in STEM. That's where the jobs of tomorrow will be."

James Brown (Executive Director of the STEM Education Coalition in Washington, D.C.)

Data from the US Bureau of Labor Statistics (BLS) support that assertion. Employment in occupations related to STEM—science, technology, engineering, and mathematics—is projected to grow to more than 9 million by 2022 [in the US alone] . . . Overall, STEM occupations are projected to grow faster than the average for all occupations.

from *STEM 101: Intro to Tomorrow's Jobs* US Bureau of Labor Statistics